Project AIR FORCE

DEFINING THE ROLE OF AIRPOWER IN JOINT MISSIONS

Glenn A. Kent
David A. Ochmanek

Prepared for the
UNITED STATES AIR FORCE

RAND

The stage is set for the emergence of a "new American way of war," in which U.S. forces are able to bring military power to bear against an enemy state quickly, comprehensively, decisively, and with minimal risk of heavy casualties. Arguably, such a transformation in U.S. military capabilities and strategy has been conceivable since the emergence of aircraft with large payloads around the time of World War II. But it has taken the emergence of new technologies and the development of new concepts for all-weather day/night surveillance and engagement, battle management, precision attack, low-observables, and other capabilities to make this new approach to warfare a reality.

Or a near reality. Today, the primary obstacles to realizing this revolution seem more budgetary and political than technical or operational. Some programs key to making this new approach a reality are being abandoned or delayed because of the press of limited resources and competing demands. In this environment, it is critical that the Air Force more clearly define the potential contributions of airpower to joint operations and the role of airpower in this emerging strategy—the "new American way of war."

The purpose of this report is to recommend a new approach to defining basic doctrine for airpower—an approach that responds to this new strategy. The approach offered here begins with a consideration of the basic characteristics of air forces and space forces, then identifies the operational capabilities of these forces, and finally lists the missions and operational objectives to which these forces can contribute. By insisting that these missions and objectives be defined from the perspective of joint operations, this approach to doctrine positions the Air Force favorably to advance the role of its forces in the competition for roles *within* missions that is envisaged in Title 10 of the U.S. Code.

This work was undertaken by the Strategy and Doctrine Program of RAND's Project AIR FORCE. It was sponsored by the Deputy Chief of Staff for Plans and Operations; Headquarters, U.S. Air Force. It should be of interest to anyone concerned about the future roles of airpower in U.S. national security.

PROJECT AIR FORCE

Project AIR FORCE, a division of RAND, is the Air Force's federally funded research and development center (FFRDC) for studies and analyses. It provides the Air Force with independent analyses of policy alternatives affecting the development, employment, combat readiness, and support of current and future aerospace forces. Research is performed in three programs: Strategy and Doctrine, Force Modernization and Employment, and Resource Management and System Acquisition.

CONTENTS

FIGURE

ACKNOWLEDGMENTS

The authors wish to thank their colleague Bill Naslund for his contributions to the ideas incorporated in this report and for his role in refining them. They are also indebted to Alan Vick, Sam Clovis, and Lt Col Paul McVinney for their thoughtful reviews of an earlier draft. Thanks are also due to Phyllis Gilmore, who edited our text, and to Shirley Birch, who brought textual order out of disorder.

Chapter One

INTRODUCTION: AIRPOWER AND THE "NEW AMERICAN WAY OF WAR"

Organizations are defined by what they do. They are distinguished from one another both by how well they do it and by other characteristics that make them unique. Like other organizations, military services seek to ensure that these distinctions are clear. Military doctrine plays a role in clarifying distinctions among the types of forces fielded by the different services.

Military doctrine has two major aspects: It *describes* what a particular form of military power can do, and it *prescribes* what that form of power should strive to become, by defining future directions that are at once consistent with the basic characteristics of the arm and yet imaginative. The purpose of this report is to suggest ways in which the U.S. Air Force can revise basic doctrine about airpower to make it a more accurate and compelling descriptive and prescriptive vehicle.

CHANGE AND BARRIERS TO CHANGE

In 1991, during Operation Desert Storm, the U.S. Air Force demonstrated that well-trained and well-equipped air forces could dominate most aspects of operations on the modern battlefield. Within a few days of the commencement of combat operations, coalition forces were able to gain the freedom to operate with near impunity over enemy territory while denying the enemy the ability to operate at all over friendly territory.[1] By laying Iraq and its ground and naval

---

[1]The sole exception being the coalition's inability to prevent Iraq from launching mobile, surface-to-surface missiles. This gap in U.S. capabilities represents a potentially serious vulnerability that must be addressed through a number of initiatives.

forces bare to observation and attack from the air, and by effectively exploiting this situation, coalition forces were able to prevent the Iraqis from regaining the initiative. Over a five-week period, the coalition forces also reduced Iraqi combat capabilities and war-making potential to such a degree that the coalition could accomplish its remaining objectives vis à vis Iraq in minimal time and with low risk of casualties.

In short, the world witnessed the fruits of decades of investment in intensive training and in new capabilities for battlefield surveillance, battle management, stealth, precision weapons, and other aspects of air operations. The result was a dramatic—indeed, revolutionary—improvement in the capabilities of air forces to locate, identify, engage, and attack a wide range of enemy assets and forces. These developments should have profound implications for the conduct of joint military operations. Yet, more than six years after Desert Storm, one finds little evidence of fundamental change in joint planning, force assessment, force structure, or resource allocation within the U.S. Department of Defense. This suggests that, despite the evidence of recent history, many in the defense community lack a clear appreciation of airpower's capabilities and potential.

There are many reasons for this, but one contributing factor is what the private sector would call poor marketing: The Air Force itself has made it easier for others to overlook, avoid, or dismiss the significance of the growth in airpower's capabilities by ineffectively articulating the case that these new capabilities should prompt a new approach to some forms of military operations. While it is hard to find an operator in the U.S. Air Force who is not highly skilled at his or her craft, it is almost as hard to find Air Force documents and statements that reflect in simple terms the true operational capabilities of modern airpower or that promulgate a doctrine that appeals to a joint audience. In short, the Air Force has not told the story of modern airpower in a clear, compelling way to the larger defense community.

A CHALLENGE FROM THE CHIEF OF STAFF

In a seminal speech, Gen Ronald Fogleman, then Chief of Staff of the U.S. Air Force, laid out what amounts to an outline of a new doctrine developed by the Air Force. He observed that the United States is "on the verge of introducing a new American way of war."[2] This new approach to warfare, he declared, was based on the emergence of modern weapon systems with extended range and increased lethality, as well as new means of surveillance, assessment, and battle management. Together, these capabilities make it possible

> to transition from a concept of annihilation and attrition warfare that places thousands of young Americans at risk in brute, force-on-force conflicts to a concept that . . . seeks to directly attack the enemy's strategic and tactical centers of gravity.

For a host of reasons, such a strategy for warfare is well suited to the national security needs of this nation. Given the right investments, U.S. leaders will, in many situations, be able to achieve national objectives quickly and decisively, without risking heavy U.S. casualties or major damage to civilian populations and infrastructure. And to the extent that such capabilities can be moved rapidly to where they are needed, the United States will be able to project power abroad rapidly and thus avoid the expense of stationing large formations overseas on a permanent basis.

Significantly, General Fogleman noted that the centers of gravity targeted in this new approach to warfare would generally include:

> the leadership elite; command and control; internal security mechanisms; war production capability; *and one, some, or all branches of its armed forces*—in short, an enemy's ability to effectively wage war [emphasis added].

Air forces, which are by their nature highly mobile and are acquiring more accurate and lethal weapons, are the forces best suited to playing the leading role in this strategy. These types of forces are increasingly capable of attacking these centers of gravity effectively. In short, we are entering an era in which the air forces of the United States have a growing potential to seize control quickly and effectively over the operations of enemy forces and assets in all other

---

[2]See *Airpower and the American Way of War*, Gen Ronald R. Fogleman, speech presented to the Air Force Association's Air Warfare Symposium, Orlando, Florida, February 15, 1996.

mediums. Because of this, as General Fogleman has observed, "airpower has . . . changed the American way of war."

MEETING THE CHALLENGE

The central premise of this report is that basic doctrine, as set forth by the U.S. Air Force in current and draft documents, is not well suited to providing a basis for understanding, developing, or advocating the role of airpower in the new American way of war General Fogleman outlined. Specifically, current doctrinal statements by the Air Force too often fail to define convincingly the missions and operational objectives in which air forces can and should play a dominant role. Further, current doctrine constrains airmen from thinking expansively about how they might play a greater role in missions traditionally relegated to surface forces.

For example, the Air Force's basic doctrine identifies its six "core competencies" as air and space superiority, global attack, precision engagement, information superiority, rapid global mobility, and agile combat support.[3] Given this list, it is difficult to find a doctrinal basis for advocating airpower on the basis of fulfilling the encompassing strategy General Fogleman envisioned.

General Fogleman sees airpower emerging as a force capable of dominating operations in virtually all dimensions of theater warfare—land, sea, air, space, and cyberspace. Yet, in its basic doctrine, the Air Force seems content to accept a minor or supporting role for airpower in countering the enemy's operations on land and sea. After all, if "air superiority" and "space superiority" are core competencies of the Air Force, then, by extension, "land superiority" and "maritime superiority" must be the provinces (competencies) of the Army and the Navy, respectively. Thus, by implication, the Air Force, through its statements of doctrine, constrains airpower from playing a more encompassing role in other domains.

Another shortcoming of current Air Force doctrine is that it runs counter to the spirit of the most important source of service responsibilities and authorities: Title 10 of the U.S. Code. Title 10—better

---

[3] *Air Force Doctrine Document 1: Air Force Basic Doctrine*, September 1977, pp. 29–35.

known as the Goldwater-Nichols Act of 1986—does not assign roles or missions to the military departments. Rather, it charges each service with the responsibility to develop military capabilities that "fulfill (to the maximum extent practicable) the current and future operational requirements of the unified and specified combatant commands."[4] This suggests that the roles the forces of any military service play are not determined on the basis of some *a priori* doctrinal claim or official demarcation but rather are "up for grabs," in the sense that these roles will be determined according to the operational relevance of the capabilities different types of forces offer. Thus, forms of military power and, by extension, the military services themselves, must *earn* roles *within* missions assigned to combatant commanders on the basis of the capabilities the types of forces each service offers. This is distinct from services *claiming* roles *and* missions for their forces on the basis of doctrinal pronouncements or listings of "core competencies."

Title 10 recognizes explicitly that the services must compete for roles in missions and do so by fielding force elements that provide the most relevant, effective, and affordable capabilities to users—the combatant commanders. *By conceptually cutting airpower out of the running for a dominant role in controlling the operations of enemy forces on the surface, current Air Force doctrine fails to position USAF forces favorably in this competition. Yet, airpower in general and USAF force elements in particular are, in reality, acquiring capabilities that will increase their ability to control the operations of enemy forces on land and at sea.*

This report seeks to remedy these shortcomings by recommending a new approach to formulating basic doctrine about airpower.[5] The approach advocated here

---

[4]See Title 10, U.S. Code, Section 3013 for the Army, 5013 for the Navy, and 8013 for the Air Force.

[5]Our approach adheres to the principle that the leaders of the Air Force assume an active and leading role in defining the doctrine that establishes the role of airpower in advancing and protecting the interests of the United States. This approach is more encompassing than defining "Air Force doctrine." We are defining doctrine developed *by* the Air Force that applies to basic characteristics of airpower, not Air Force doctrine that applies only to forces fielded by the USAF.

- supports an expansive concept of the potential roles of airpower in future operations
- is consistent with the letter and spirit of Title 10
- lends itself readily to operationally oriented assessments and comparisons of the capabilities and limitations of all types of forces, including those the Air Force provides.

In combination, these features of the approach outlined here make it far better suited than existing doctrine to supporting the role of airpower in a new American way of war.

Chapter Two

THE SOURCE OF BASIC DOCTRINE: FUNDAMENTAL CHARACTERISTICS OF AIRPOWER

Basic doctrine about airpower should begin with an identification of the inherent and fundamental characteristics of air forces. Doctrine then spells out the implications of these fundamental characteristics for military strategy and operations.

The fundamental characteristic of air forces is that their basic items of equipment (aircraft) operate in the medium of the air. As stated in the Air Force Dictionary[1]

> Basic air doctrine deals with the phenomenon of flight, with the new relationships that exist as a result of hitherto unrealized speeds, range, mobility, and flexibility, and their application to the principles of war . . . .

So far, so good. But various iterations of basic doctrine about airpower have had a mixed record in terms of the degree to which they have provided a basis for advocating important roles for air forces in joint operations.

APPROACHES TO USAF DOCTRINE: A COMPARISON

Building on earlier versions of basic doctrine, Air Force Manual 1-1 of 1975 included a straightforward statement about the fundamental characteristics and capabilities of aerospace forces:

---

[1]Woodford Agee Heflin, ed., *The United States Air Force Dictionary*, Washington, D.C.: Air University Press, 1956, p. 74.

2-1. Aerospace. The region above the earth's surface permits largely unencumbered access to any point on or above the earth. This provides opportunity for direct application of aerospace power against all elements of an enemy's resources regardless of their location.

2-1. Characteristics and Capabilities of Aerospace Forces. The freedom of operations permitted in aerospace allows aerospace forces to exploit speed, range, altitude, and maneuverability characteristics to a degree not possible by other forces. Unimpeded by natural barriers imposed by land and water masses, aerospace forces can conduct operations rapidly, over great distances in any direction, and enjoy multidimensional maneuvering within the medium of aerospace. Certain distinctive capabilities have evolved through exploitation of these characteristics. Chief among them are:

a. Flexibility

b. Responsiveness

c. Survivability

d. Surveillance.

As recently as 1990, at least one official Air Force document continued to hew to this basic formulation:

> This paper provides a perspective on how the unique characteristics of the Air Force—speed, range, flexibility, precision, and lethality—can contribute to underwriting U.S. national security needs in the evolving world order. It also challenges Air Force members, and others in the defense establishment, to think about how we as a nation can best address the role of military forces for the future. And finally, the concepts outlined here . . . provide a framework to conduct future Air Force planning.[2]

This approach to defining the characteristics and capabilities of airpower is noteworthy because it is outward looking and expansive: It communicates a recognition that airpower has the potential to support a broad range of the security needs of the United States. It

---

[2]Donald Rice, Secretary of the Air Force, *The Air Force and National Security*, 1990, p. 1.

highlights, in clear terms, that which is unique and valuable about air and space forces.

Unfortunately, one can also find authoritative Air Force documents that embody a more constraining approach. Perhaps the most egregious example is found in Air Force Manual 1-1 of 1984, which states that:

- The basic objective of land forces is to win the land battle.

- The basic objective of naval forces is to win the naval battle.

- The basic objective of aerospace forces is to win the aerospace battle.

This formulation constrains airpower to, at most, a subsidiary role in defeating enemy surface forces and thus runs directly counter to General Fogleman's conception of a growing role for airpower in warfare.

Air Force Manual 1-1 of 1992 dropped this flawed language but introduced new terms that are also problematical:

Aerospace forces perform four basic roles: aerospace control, force application, force enhancement, and force support.

The 1992 revision of AFM 1-1 does provide an opening for airmen to conceive of a growing role for airpower in joint operations. That opening is something called "force application," which is said to include, among other things, interdiction. Because interdiction encompasses attacks on forces operating in other mediums—namely, land and sea—it opens up a means within doctrine for airpower to play some role in "the land battle" and "the naval battle."

Nevertheless, the term "force application" lacks the simple directness and ringing clarity of earlier statements. Compare it, for example, to an eloquent statement by Maj Gen Frank M. Andrews in 1938:

The airplane is the only weapon which can engage with equal facility, land, sea, and other forces . . . .[3]

---

[3]Quoted in Air Force Manual 1-1, 1984, p. 3-1.

This statement suggests that General Andrews, rather than constraining himself to a narrow view of what airpower could accomplish, envisioned the potential capabilities of airpower in all mediums of warfare. Now that the capabilities of airpower have caught up with this vision, the Air Force ought to promulgate a doctrine that, at a minimum, accurately describes this new reality.

A WINNING SYLLOGISM

This review of past statements about the nature and role of airpower, while admittedly selective and somewhat cursory, suggests two conclusions:

- First, statements about the role of airpower by military leaders (past and present) were more encompassing and more compelling than statements found in the last two versions of AFM 1-1.

- Second, inherent in the better statements is a three-part syllogism:

  — Air forces and space forces, by operating in the mediums of air and space, possess the *basic characteristics* of speed, range, maneuverability, perspective, and mobility.

  — By exploiting these basic characteristics, modern airpower provides operational *capabilities* fundamental to the conduct of war, such as mass, maneuver, and situational awareness.

  — These capabilities portend a major and growing *role* for airpower in achieving important operational objectives and in protecting and advancing the interests of the United States.

Approaching the formulation of doctrine in this manner—from basic characteristics, to capabilities, to roles—yields a conceptual framework that is both logically coherent and consistent with Title 10's approach (that is, that forces earn roles in missions by virtue of the capabilities they bring to combatant commanders). Such an approach positions air forces (and, hence, the Air Force) favorably for the competition envisaged in Title 10.

The following formulation fleshes out this approach and defines the nature and roles of air and space forces:

- Air forces and space forces operate in the mediums of air and space.

- The mediums of air and space are continuous around the globe, have no boundaries, are above the mediums of land and sea, permit observation of operations in these other mediums, and provide free access to any point on or above the earth.

- Forces operating in the air possess the inherent characteristics of speed, range, maneuverability, perspective, and mobility of large payloads.[4]

- These inherent characteristics yield operational capabilities central to the conduct of war (rapid and global projection of power, responsiveness, surprise, mass, dispersion, maneuver, and situational awareness). Air forces and space forces thereby provide the opportunity to gain perspective over the entire battle space and to apply power directly against all elements of an enemy's resources, regardless of their location.

- Recent advances in technology have enabled new engagement systems and weapons that yield order-of-magnitude increases in the effectiveness and lethality of weapons delivered by air forces.

- By virtue of the nature of their mediums, the inherent characteristics of forces operating in those mediums, and the continuing enhancements to the effectiveness and lethality of these forces, *air forces and space forces have the potential to dominate the conduct of enemy operations in all mediums—operations on the land, at sea, in the air, and through space and operations by enemy leaders in exercising sovereignty over their country.*

- The capabilities offered by modern airpower are especially well suited to meeting many of the demands of U.S. military strategy. Specifically, they allow operations characterized by rapid and long-range projection of effective military power.

---

[4]Space-based assets share many of the characteristics of air forces. However, the air and space mediums are distinct and call for different types of vehicles. From the perspective of this paper, which focuses on the capabilities of and doctrine for air forces, spaced-based assets provide important capabilities supplementary to those of air (and terrestrial) forces.

- It follows that airpower can and should play an increasingly significant role in providing the operational capabilities directed toward controlling all enemy operations, in achieving national security objectives, and in advancing and protecting the security interests of the United States.

We do not claim that airpower is today capable of dominating operations in all mediums or in all circumstances. Neither do we believe that the U.S. Air Force should strive for a dominant role in all aspects of military operations. For example, air forces cannot compel a dug-in enemy ground force to move, cannot seize and hold territory, and cannot effectively patrol urban areas. General Fogleman again struck the right note here when he said, in the same speech, that "warfare today, and in the future, will be joint warfare." Our objective is to advance a doctrinal framework that removes self-imposed limitations on thinking about the potential roles of air and space forces.

In addition, the formulation advocated here provides a conceptual grounding for conceiving new approaches to employing air forces and space forces and for identifying priorities for the allocation of resources to modernize those forces. To support the latter activities, it is necessary to disaggregate these broad statements of the role of airpower into operational-level components. This is the subject of the next chapter.

A WORD ABOUT "CORE COMPETENCIES"

The concept of "core competencies" has acquired some currency recently and has been used in many ways. If this term is to have any value in the realm of doctrine, it should be applied to those qualities of different types of forces that are unique and enduring. If we adopt this standard for the use of the term, the core competency of air and space forces—that which sets them apart from other types of forces—is that they are indeed air and space forces: These forces, by nature, fly through the air and traverse space—the mediums of choice for accomplishing many critical operational tasks.

Used in this way, the concept of a core competency reinforces the approach to doctrine advocated above. That is, the core competency of air and space forces—their ability to traverse air and space—gives them inherent characteristics of speed, range, mobility, and perspective. These inherent characteristics, in turn, make it possible for air and space forces to possess the fundamental capabilities of projection, responsiveness, maneuver, mass, and situation awareness. Attempting to define the core competencies of these forces in more specific terms than this (for example, "air and space superiority," "information superiority") runs the risk of overspecifying and, hence, limiting the potential employment of air and space forces. It would be awkward for the Air Force to proclaim that its units possessed a core competency of "land superiority," even though these units can, in fact, dominate operations on the land in many cases.

An approach to defining these forces in terms of inherent characteristics and capabilities seems preferable to "core competencies" for another reason: The capabilities of a force or a force element can be measured and compared to those of other forces. Core competencies, by contrast, are intended to be unique to each service and, hence, by definition, do not lend themselves well to comparisons. As defined thus far, core competencies are also too general to permit meaningful comparison. If one believes that air forces compare favorably to other types of forces in terms of their capability to perform key operational tasks, it follows that the Air Force should strive to define these forces in terms that invite, rather than avoid, quantification and comparison.

Chapter Three

TOWARD A DEFINITION OF ROLES
WITHIN MISSIONS

Past statements of basic doctrine by the Air Force traditionally include a listing of "Air Force missions," such as interdiction, close air support, defensive counterair, and others. These lists have not enjoyed a great deal of saliency outside of the Air Force, mainly because they are not really "Air Force missions." Missions are assigned to combatant commanders, not services. They are carried out by force elements provided by the services, usually in a joint operation. For example, Air Force and Navy aircraft may carry out interdiction sorties; Air Force, Army, and Marine aircraft may provide close support to friendly ground forces; and all four services field forces capable of shooting down enemy aircraft in flight. In light of this reality, it makes little sense for the Air Force to try to lay claim to a set of missions through its own doctrinal statements.

General Fogleman recognized this when, in 1995, he charged the Air Force's long-range planning activity to "define the roles of air and space forces in joint missions." Since missions are inherently joint and since the Air Force should, in principle, be poised to demonstrate growing capabilities across the full range of joint missions, we recommend that the Air Force adopt an approach to defining the capabilities of air and space forces that is cast in terms of joint missions and expressed at the operational level. This is quite distinct from claiming a subset of joint missions as "belonging" to the Air Force.[1]

---

[1]Of course, the other services could use a list identical to the one we propose here.

This begs the question of what joint missions are. Webster defines a mission as "a specific task with which a person or group is charged." Inherent in the concept of mission is an objective—something that is to be achieved. With this in mind, we offer the following list of general missions that are assigned or that could be assigned to U.S. combatant commanders in peace and war:

- **Strengthen stability and deter aggression** in key regions through operations and interactions in peacetime.

- **Resolve crises** by, for example, providing humanitarian relief, enforcing peace agreements, protecting and evacuating civilians, rescuing hostages, conducting punitive strikes, or intervening against hostile regimes.

- **Win "cold wars."** By this is meant the military dimensions of a long-term political, economic, and military strategy aimed at isolating and exerting pressure upon an adversary state. The Soviet Union was the primary target of such a strategy during the Cold War. Today, North Korea, Iraq, Libya, and Cuba are primary among the states with which the United States is waging a cold war.

- **Counter weapons of mass destruction.** This encompasses efforts to impede the proliferation of nuclear, chemical, and biological weapons, as well as their delivery vehicles; to deter their use; and to prevent them from being used (or used effectively), through counterforce attacks and the employment of active and passive defenses.

- **Defeat large-scale aggression and compel surrender.** This mission is, properly, the focus of the bulk of the United States' defense planning and defense resources. The ability to project large-scale military forces over long distances is the defining feature of U.S. military forces in the post–Cold War era, and it is this ability that makes the U.S. alliance structure viable. Accordingly, we focus on this mission in the remainder of this document.

Each of these broad missions can be disaggregated into its constituent components. In the case of the key mission of defeating large-scale aggression, the constituent parts are the elements of an overall theater campaign. Not all of these objectives would neces-sarily be relevant in every conflict, and they would not have to be pursued in a strictly sequential fashion. (Indeed, there is great value in being able to attack multiple objectives simultaneously to maxi-mize the shock value of one's operations and minimize the enemy's ability to adapt effectively. Such a "parallel" approach to attacking enemy assets would, presumably, offer the best chance of compelling the enemy to surrender quickly.)

Assuming that the enemy initiates a major conflict through offensive action, a generic theater campaign can be thought of as having two major phases. In the first phase U.S. and allied forces would strive to wrest from the enemy the initiative in all aspects of military opera-tions within the theater. As this was accomplished, the focus would shift to destroying the enemy's capacity to wage war and compelling a surrender.[2] Key objectives within each of these phases are listed below and are displayed graphically in Figure 1.

• **Phase I: Gain Control of Military Operations**

—Dominate operations in the air

—Dominate maritime operations

—Halt invading land forces short of critical objectives

—Defeat weapons of mass destruction and their delivery means

  • Destroy or deny access to stocks of weapons of mass destruction and delivery means

  • Destroy aircraft, cruise missiles, and ballistic missiles in flight

---

[2]This conception of a theater campaign is somewhat different from the more com-monly used three-phased approach, in which U.S. and allied forces would seek to halt the enemy invasion, reduce the enemy's warfighting capacity while building up their own in the theater, and then compel the enemy's forces to withdraw from captured territory through a counteroffensive. We prefer our approach because there is more to the opening phase of a conflict than simply halting aggression and because it is not a foregone conclusion that a combined arms counteroffensive will always be necessary to compel surrender. (Both approaches envisage a post-war stabilization phase of indefinite duration, but operations in this phase are subsumed within our framework under the mission of strengthening stability.)

RAND *MR927-1*

Gain Control	Compel Surrender
Dominate air operations	Degrade/destroy warmaking infrastructure
Dominate maritime operations	Isolate, demoralize, destroy fielded forces
Halt the invasion	Evict enemy forces from captured territory
Deter/prevent WMD use	Seize and hold ground

Figure 1—Key Objectives in Theater Warfare

- Protect U.S. and allied assets, facilities, and personnel from the effects of weapons of mass destruction.

- Pose credible threats of unacceptable damage or retaliation for use of weapons of mass destruction.

- **Phase II: Continue Operational Dominance and Compel Surrender**

  —Degrade and destroy war-related industries and facilities that support the conduct of the war

  - Especially, neutralize facilities that support the enemy's leadership in maintaining control over military operations and in exercising sovereignty over the nation

  —Further isolate, destroy, and demoralize enemy military forces

  —Evict enemy forces from captured territory, if necessary

  —Seize and hold ground.

This framework encompassing joint missions and objectives provides a means by which the Air Force can describe the capabilities of air forces and space forces and then quantify these capabilities in analyses of representative campaigns. A basic doctrine that set forth such a framework would help position the Air Force favorably to compare and debate the utility of air and space forces in meeting the needs of the future.

Chapter Four

ADVANTAGES OF A NEW APPROACH TO BASIC DOCTRINE

The leadership of the Air Force has issued a clear and compelling challenge to strive to realize a vision of a future in which U.S. leaders will be able to achieve national objectives quickly and decisively, without risking heavy U.S. casualties or major unwanted collateral damage. Properly supported and equipped, modern air and space forces are well endowed to play the dominant role in such an approach to major theater conflicts. The technological and operational components of this revolutionary new way of war appear to be largely in place. But if the nation is to reap the advantages of this new approach to warfare, airpower must continue to advance toward fulfilling its full potential. One important step in achieving this will be a clear statement by the U.S. Air Force about the roles of air and space forces in future conflicts. This statement must free the minds of airmen to think expansively about the inherent qualities of air and space forces and their potential capabilities.

The "joint mission" approach advocated here meets these requirements. It

- Builds upon the most fundamental aspects of air and space forces—their ability to transit the seamless mediums of air and space

- Has immediate saliency in the joint community and with the users of forces provided by the services—the combatant commanders

- Is neutral with regard to the role that each service plays in any stated mission

- Positions the Air Force favorably to advance concepts under which air and space forces play greater roles in key missions

- Provides a basis for assessing the capabilities and limitations of various force elements, such that the capabilities of forces can be compared at the operational level and results can be readily explained to audiences outside of the Air Force.

Accordingly, we believe that the leadership of the Air Force should consider adopting this new approach as the conceptual basis of a revised statement of basic doctrine. Doing so would contribute greatly to the realization of the vision in which airpower plays an increasingly dominant role in the American way of war.